CHAPTER 1

Introduction

Artificial Intelligence (AI) is a branch of computer science that focuses on creating machines that can perform tasks that would typically require human intelligence, such as learning, decision-making, perception, and natural language processing. AI is a rapidly evolving field that has the potential to transform numerous industries and applications, from healthcare to finance to transportation.

AI can be classified into two main categories: narrow or weak AI, and general or strong AI. Narrow AI is designed to perform specific tasks, such as image recognition or natural language processing, and is the most common type of AI used in commercial applications today. General AI, on the other hand, is designed to perform any intellectual task that a human can do and is still largely a theoretical concept.

There are several different approaches to developing AI, including rule-based systems, evolutionary algorithms, and machine learning. Machine learning is one of the most popular and successful approaches to AI and involves training algorithms to learn from data without being explicitly programmed. This approach has led to significant advances in areas such as computer vision, natural language processing, and autonomous systems.

AI can also be categorized into three main types based on their level of awareness and memory: reactive machines, limited memory, and self-aware AI. Reactive machines are designed to react to specific inputs and produce specific outputs without any memory or ability to learn from past experiences. Limited memory AI systems can learn from past experiences and use that information to make decisions in the present. Self-aware AI is still largely theoretical and involves machines that have a sense of consciousness and can reason about their own existence.

Despite the numerous benefits of AI, there are also important ethical and societal considerations that must be taken into account, such as the potential for job displacement, biases and discrimination perpetuated by AI systems, and the implications of AI-powered decision-making. As AI technologies continue to evolve, it is essential to approach their development and deployment with responsibility and caution.

Artificial Intelligence (AI) is a rapidly growing field that has the potential to transform many aspects of our lives. At its core, AI is focused on creating machines that can perform tasks that would typically require human intelligence, such as learning, decision-making, perception, and natural language processing. This chapter will explore the different types of AI, the approaches used to develop AI, and the benefits and drawbacks of AI.

Types of AI

There are two main types of AI: narrow or weak AI, and general or strong AI. Narrow AI is designed to perform specific tasks and is the most common type of AI used in commercial applications today. Examples of narrow AI include image recognition, natural language processing, and speech recognition systems. In contrast, general AI is designed to perform any intellectual task that a human can do and is still largely a theoretical concept.

AI can also be categorized into three main types based on their level of awareness and memory: reactive machines, limited memory, and self-aware AI. Reactive machines are designed to react to specific inputs and produce specific outputs without any memory or ability to learn from past experiences. Limited memory AI systems can learn from past experiences and use that information to make decisions in the present. Self-aware AI is still largely theoretical and involves machines that have a sense of consciousness and can reason about their own existence.

Approaches to Developing AI

There are several different approaches to developing AI, including rule-based systems, evolutionary algorithms, and machine learning. Rule-based systems involve programming machines with a set of rules to follow when making decisions. Evolutionary algorithms involve using techniques inspired by natural selection to optimize machine performance. Machine learning, on the other hand, involves training algorithms to learn from data without being explicitly programmed.

Machine learning is one of the most popular and successful

approaches to AI, and it has led to significant advances in areas such as computer vision, natural language processing, and autonomous systems. There are three main types of machine learning: supervised learning, unsupervised learning, and reinforcement learning. In supervised learning, machines are trained on a set of labelled examples to learn how to classify new data. In unsupervised learning, machines learn to identify patterns in data without any labels. In reinforcement learning, machines learn by trial and error and receive feedback in the form of rewards or punishments.

Benefits and Drawbacks of AI

The benefits of AI are numerous, including increased efficiency and productivity, improved accuracy, and the ability to perform tasks that are too dangerous or difficult for humans. AI can also lead to new and innovative solutions to complex problems, such as disease diagnosis and drug development.

However, there are also significant drawbacks to AI that must be considered. One of the most significant concerns is the potential for job displacement, as machines take over tasks previously performed by humans. There also the risk of biases and discrimination perpetuated by AI systems, as algorithms learn from historical data that may contain biases. Additionally, there are concerns about the implications of AI-powered decision-making, particularly in areas such as criminal justice and healthcare.

Conclusion

AI is a rapidly evolving field with enormous potential to transform many aspects of our lives. However, it is essential to approach the development and deployment of AI with caution, considering the ethical and societal implications of these technologies. By understanding the different types of AI, approaches to developing AI, and the benefits and drawbacks of AI, we can make informed decisions about how to leverage this technology for the betterment of society.

History of AI

The history of artificial intelligence (AI) dates back to ancient times, with the earliest recorded attempts to create intelligent machines dating back to Greek myths about mechanical men. However, it was not until the 20th century that AI began to take shape as a scientific discipline.

In the 1950s and 1960s, the pioneers of AI, including John McCarthy, Marvin Minsky, and Claude Shannon, began to develop the first AI programs and concepts. These early AI systems were rule-based and relied on sets of programmed rules to make decisions and perform tasks.

In the 1960s and 1970s, AI experienced significant growth and innovation, with researchers exploring new approaches to developing intelligent machines, such as machine learning and natural language processing. The development of the first AI programs capable of playing chess and other games marked a significant milestone in AI research.

However, by the 1980s, progress in AI research had slowed, and the field entered a period of reduced funding and interest. This period, known as the "AI winter," lasted until the late 1990s, when advances in computing power and the emergence of the internet sparked renewed interest in AI.

In recent decades, AI has experienced explosive growth and development, with breakthroughs in areas such as computer vision, natural language processing, and machine learning. The development of powerful machine learning algorithms and large datasets has led to significant advances in areas such as image recognition, speech recognition, and autonomous systems.

Today, AI technologies are ubiquitous and have applications in a wide range of industries and fields, from healthcare to finance to transportation. With the continued evolution of AI, the potential for AI-powered innovation and transformation remains vast, and the field is likely to continue to evolve and expand in the years to come.

Key concepts

Artificial intelligence (AI) is a complex field with many key concepts that are essential to understanding its applications and potential impacts. Some of the most important key concepts in AI include:

Machine learning: A subset of AI that involves training algorithms to learn from data without being explicitly programmed. Machine learning has been used to make significant advances in areas such as image recognition, natural language processing, and autonomous systems.

Natural language processing (NLP): A field of AI that focuses on enabling machines to understand and interact with human language. NLP has applications in areas such as virtual assistants, chatbots, and language translation.

Computer vision: A field of AI that involves enabling machines to interpret and understand visual information from the world around them. Computer vision has applications in areas such as autonomous vehicles, facial recognition, and security systems.

Deep learning: A subset of machine learning that involves training artificial neural networks with many layers. Deep learning has been used to achieve breakthroughs in areas such as speech recognition and image recognition.

Neural networks: Algorithms that are modelled after the structure and function of the human brain. Neural networks are used in many areas of AI, including image recognition and natural language processing.

Robotics: The development and use of robots that can perform tasks autonomously or with minimal human intervention. Robotics has applications in areas such as manufacturing, healthcare, and transportation.

Ethics and governance: The ethical and societal implications of AI, including concerns around job displacement, biases and discrimination perpetuated by AI systems, and the implications of AI-powered decision-making. It is essential to approach the development and deployment of AI with responsibility and caution.

Understanding these key concepts is crucial to developing and deploying AI systems effectively and responsibly. As AI technologies continue to evolve, it is important to stay abreast of the latest developments and ethical considerations in the field.

Current state of development

The current state of development in artificial intelligence (AI) is rapidly advancing, with new breakthroughs and applications emerging regularly. Some of the key areas of progress in AI include:

Machine learning: The development of more advanced machine learning algorithms and frameworks has enabled significant progress in applications such as natural language processing, computer vision, and autonomous systems.

Deep learning: The use of deep learning algorithms, which are modeled after the structure and function of the human brain, has led to breakthroughs in areas such as speech recognition and image recognition.

Robotics: Advances in robotics technology have led to the development of more capable and versatile robots for a wide range of applications, from manufacturing to healthcare to transportation.

Autonomous systems: The development of autonomous systems, such as self-driving cars and drones, is rapidly advancing and has the potential to transform transportation and logistics.

Healthcare: AI is being used to develop new diagnostic tools and personalized treatment plans, and is also being used to analyse medical data to identify new patterns and insights.

Finance: AI is being used to develop more sophisticated and accurate financial models, and to identify patterns and trends in financial data that can inform investment strategies.

Natural language processing: Advances in natural language processing have led to the development of virtual assistants and chatbots that can interact with humans in a more natural and intuitive way.

Despite these advances, there are still many challenges and limitations in the field of AI. One of the biggest challenges is ensuring that AI systems are ethical and unbiased, and do not perpetuate or amplify existing biases and discrimination. Additionally, there are concerns around the potential for job displacement as AI becomes more prevalent in the workforce. As the field of AI continues to develop, it will be important to address these challenges and ensure that the benefits of AI are distributed equitably across society.

Different Types Of AI

There are several different types of artificial intelligence (AI), each with its own capabilities and limitations. Here are some of the most commonly recognized types:

Reactive machines: These are the most basic types of AI systems, which only react to specific inputs without any memory or ability to learn from past experiences. Examples include chess-playing programs and voice assistants like Siri and Alexa.

Limited memory AI: These systems are designed to learn from past experiences and use that information to make decisions in the present. Examples include self-driving cars, which use data from sensors to make decisions in real-time.

Theory of mind AI: These systems are still largely theoretical, but involve machines that can understand and interpret human emotions and thoughts. This type of AI is being explored in applications such as virtual assistants and customer service chatbots.

Self-aware AI: This is the most advanced type of AI, which involves machines that have a sense of consciousness and can reason about their own existence. This type of AI is still largely theoretical, and is the subject of much philosophical debate.

Narrow or weak AI: This type of AI is designed to perform specific tasks, such as image recognition or natural language processing. Most of the AI applications in use today fall into this category.

General or strong AI: This type of AI is designed to perform any intellectual task that a human can do, and is still largely a theoretical concept. This type of AI would require the ability to reason, plan, and understand language at a human-like level.

The development of AI is still in its early stages, and there is ongoing research and experimentation in all of these different types. As AI becomes more advanced, it is likely that new categories and sub-categories will emerge.

All of these subfields of AI rely on large amounts of data to train algorithms and improve their accuracy. As more data becomes available, machine learning, natural language processing, and computer vision are likely to continue to advance and have a significant impact on industries such as healthcare, finance, and transportation. However, there are also important ethical considerations that must be taken into account, such as privacy and biases in data and algorithms.

CHAPTER 2:

The Pros of AI in the Workplace

Artificial intelligence (AI) has the potential to transform the workplace in numerous ways, offering several advantages and benefits for businesses and employees alike. In this chapter, we will explore some of the key advantages of AI in the workplace.

Increased Efficiency: One of the primary advantages of AI is that it can automate repetitive and time-consuming tasks, allowing employees to focus on higher-level work. AI can also perform tasks faster and more accurately than humans, leading to increased productivity and efficiency in the workplace.

Improved Accuracy: AI can be trained to make more accurate predictions and decisions than humans, particularly in complex or data-driven environments. This can lead to better decision-making and improved outcomes in areas such as healthcare, finance, and logistics.

Cost Savings: Implementing AI in the workplace can lead to significant cost savings, particularly in industries such as manufacturing and logistics where automation can reduce the need for human labour. AI can also help businesses identify areas of waste or inefficiency and make recommendations for improvement.

Enhanced Customer Experience: AI can be used to improve the customer experience by providing personalized recommendations, chatbots for customer service, and more efficient and accurate processing of customer data.

Improved Safety: AI can be used to monitor and analyse data to identify potential safety hazards in the workplace, allowing businesses to take proactive measures to prevent accidents and injuries.

Innovation: AI can drive innovation and enable businesses to develop new products, services, and business models. For example, AI-powered data analysis can provide businesses with insights into customer behaviour and preferences, allowing them to develop more targeted marketing strategies and product offerings.

In conclusion, AI offers several potential benefits for businesses and employees alike, including increased efficiency, improved accuracy, cost savings, enhanced customer experience, improved safety, and innovation. As AI technologies continue to evolve, it is important for businesses to consider how they can incorporate these technologies into their operations to remain competitive and meet the needs of their customers.

CHAPTER 3:

The Cons of AI in the Workplace

While there are numerous advantages to incorporating artificial intelligence (AI) into the workplace, there are also several potential disadvantages and drawbacks that must be considered. In this chapter, we will explore some of the key cons of AI in the workplace.

Job Displacement: One of the most significant concerns with AI is the potential for job displacement. As machines become more capable of performing tasks that were once performed by humans, there is a risk that many jobs will become automated. This can lead to job loss and the need for workers to develop new skills in order to remain relevant in the workforce.

Bias and Discrimination: Another potential issue with AI is the risk of bias and discrimination perpetuated by AI systems. AI systems can learn from data, and if that data is biased or discriminatory, the AI system may also exhibit those biases. This can lead to unfair treatment of certain groups, such as minority or marginalized communities.

Lack of Transparency: AI systems can be complex and difficult to understand, particularly for non-experts. This can make it challenging to assess how decisions are being made and to identify potential biases or errors in the system.

Overreliance on Technology: While AI can offer numerous benefits, overreliance on technology can also be problematic. If a business becomes too dependent on AI systems and those systems fail, it can lead to significant disruptions and delays in operations.

Security Risks: AI systems can also pose security risks, particularly if they are connected to the internet or other networks. Hackers can potentially gain access to sensitive information or even take control of AI systems.

Ethical Concerns: Finally, there are several ethical concerns associated with AI, particularly around the use of AI in areas such as military or law enforcement. For example, there is a risk that autonomous weapons could be developed, leading to a loss of human control and potentially catastrophic consequences.

In conclusion, while there are many advantages to incorporating AI into the workplace, there are also several potential drawbacks and concerns that must be considered. It is important for businesses to carefully weigh the pros and cons of AI and to take steps to mitigate potential risks and challenges. This may include developing ethical guidelines for the use of AI, investing in employee training and re-skilling programs, and ensuring transparency and accountability in AI decision-making processes.

CHAPTER 4:

The Impact of AI on Employment

The integration of artificial intelligence (AI) into the workplace has the potential to significantly impact employment across numerous industries. In this chapter, we will explore some of the key ways in which AI may impact employment.

Job Displacement: As mentioned earlier, one of the most significant concerns with AI is the potential for job displacement. As AI systems become more capable of performing tasks that were once performed by humans, there is a risk that many jobs will become automated, leading to job loss and the need for workers to develop new skills in order to remain relevant in the workforce.

New Job Creation: While some jobs may become automated, the development and implementation of AI systems can also lead to the creation of new jobs. These may include roles such as AI developers, data scientists, and machine learning engineers.

Changes in Job Requirements: As AI becomes more integrated into the workplace, the nature of certain jobs may also change. For example, customer service representatives may need to develop new skills in order to interact with AI-powered chatbots, while truck drivers may need to learn how to operate autonomous vehicles.

Increased Productivity: AI systems have the potential to significantly increase productivity in the workplace, allowing businesses to accomplish more with fewer resources. This can lead to cost savings and improved efficiency, which can ultimately benefit both businesses and workers.

Skills Gap: The integration of AI into the workplace may also exacerbate existing skills gaps, as workers who lack the necessary skills to work with AI systems may be at a disadvantage in the job market. This highlights the importance of investing in education and training programs to ensure that workers have the skills they need to succeed in an AI-driven workforce.

Ethical Considerations: Finally, the impact of AI on employment also raises important ethical considerations. For example, there is a risk that AI systems could be used to monitor or even control workers, potentially leading to violations of workers' rights and privacy.

In conclusion, while the impact of AI on employment is complex and multifaceted, it is clear that AI has the potential to significantly impact the workforce in numerous ways. As such, it is essential for businesses and policymakers to carefully consider the potential impacts of AI on employment and to take steps to mitigate potential risks and challenges. This may include investing in education and training programs, developing ethical guidelines for the use of AI in the workplace, and working to ensure that the benefits of AI are shared fairly and equitably among all workers.

DR. IRFAN ASHRAF

CHAPTER 5:

The Future of AI in the Workplace

As AI technology continues to advance and become more widely adopted in various industries, the future of AI in the workplace is both exciting and uncertain. Here are some potential developments and challenges that may shape the future of AI in the workplace:

Increased automation: AI-powered automation is likely to continue to replace human labor in certain industries and job functions. This could lead to increased efficiency and productivity, but may also result in job displacement and a need for workers to develop new skills and adapt to new roles.

New job opportunities: While some jobs may be automated or eliminated, AI is also likely to create new job opportunities in areas such as data analysis, AI development, and human-machine interaction. Workers may need to acquire new skills and training to take advantage of these opportunities.

Enhanced workplace safety: AI can improve workplace safety by identifying potential hazards, monitoring worker behaviour, and providing real-time feedback to prevent accidents and injuries.

Ethical concerns: As AI becomes more sophisticated and powerful, there are concerns about the potential misuse of AI in the workplace, such as discrimination, invasion of privacy, and biased decision-making. It will be important for companies and policymakers to address these ethical concerns and ensure that AI is used responsibly.

Integration with human workers: AI is not intended to replace human workers entirely, but rather to work alongside them to enhance productivity and efficiency. The challenge will be to integrate AI into the workplace in a way that supports and augments human workers without diminishing their value or agency.

Continued innovation: AI technology is constantly evolving, and the future of AI in the workplace will depend on continued innovation and development. New breakthroughs in areas such as natural language processing, machine learning, and robotics could transform the workplace and create new opportunities and challenges for workers and companies.

Overall, the future of AI in the workplace is likely to be characterized by both benefits and challenges. While AI has the potential to enhance productivity, safety, and innovation, it is important to address ethical concerns and ensure that workers are equipped with the skills and training needed to adapt to new roles and job functions. As AI technology continues to evolve, it will be important for companies and policymakers to carefully consider its impact on the workforce and society as a whole.

CHAPTER 6

Conclusion

In conclusion, the rise of artificial intelligence (AI) in the workplace is a complex and rapidly evolving phenomenon that has the potential to transform numerous industries and applications. AI technologies have the ability to enhance productivity, efficiency, and workplace safety, while also creating new job opportunities and innovations. However, there are also concerns about the potential impact of AI on employment, ethical considerations, and the need for workers to acquire new skills and adapt to new roles.

As AI technology continues to advance and become more widely adopted, it is important for companies and policymakers to approach its development and deployment with responsibility and caution. This includes addressing ethical concerns, ensuring that workers have the skills and training needed to adapt to new roles, and integrating AI into the workplace in a way that supports and augments human workers.

Ultimately, the future of AI in the workplace will depend on continued innovation, responsible development, and a thoughtful approach to its impact on the workforce and society as a whole. By harnessing the potential of AI while also addressing its challenges and limitations, we can create a future in which AI technology enhances the human experience and contributes to a more productive, efficient, and equitable workplace.

REFERENCES

Here are some references that were used to inform this book on the rise of artificial intelligence in the workplace:

Russell, S. J., & Norvig, P. (2010). Artificial intelligence: a modern approach (3rd ed.). Upper Saddle River, N.J.: Prentice Hall.

Brynjolfsson, E., & McAfee, A. (2014). The second machine age: work, progress, and prosperity in a time of brilliant technologies. New York: W.W. Norton & Company.

Davenport, T. H., & Kirby, J. (2015). Beyond automation: strategies for remaining gainfully employed in an era of very smart machines. Harvard Business Review, 93(6), 58-65.

Manyika, J., Chui, M., Miremadi, M., Bughin, J., George, K., Willmott, P., & Dewhurst, M. (2017). A future that works: automation, employment, and productivity. McKinsey Global Institute.

Simonite, T. (2018). The AI boom is happening all over the world, and it's accelerating quickly. MIT Technology Review.

Holmstrom, J., & Rantala, T. (2020). Automation, skills and the future of work: The effects of technological change on education and training needs. OECD Education Working Papers, No. 214.

Susskind, R., & Susskind, D. (2015). The future of the professions: how technology will transform the work of human experts. Oxford University Press.

Ford, M. (2015). Rise of the robots: technology and the threat of a jobless future. Basic Books.